Some Basics of Chemistry Useful in Redox Processes

By Malika Ammam, PhD

Copyright© 2017 Malika Ammam. All rights reserved.

Discount Offers

5% OFF of the book price for purchases of 1-5 books.

8% OFF of the book price for purchases of more than 5 books.

To receive the discount money, send your request through https://www.malika-ammam.com/ with your order details and PayPal account. Make sure that your order details (amazon or other sites) passed the 30 days return policy.

Thank you,

Introduction

As a teacher of physical chemistry, I noticed that students, even in advanced classes, have difficulties in understanding the basics of redox chemistry. In this Section 1, I attempted to summarize some fundamental principles of chemistry often found later in redox processes. Students have often difficulties in linking CHEMISTRY to REDOX CHEMISTRY (or ELECTROCHEMISTRY), thinking that both are separate from each other. Chemistry concepts, such as solvents, electrolytes, ionic conductivity, acidity, alkalinity (basicity), *pH*, as well as some thermodynamics and kinetics basics are all concepts that could be found in redox chemistry. To further clarify the discussed concepts, numerous questions and problems with detailed answers are provided. Most of these questions are formulated by students like you. I believe that this Section 1 will greatly help students with levels varying from high school to advanced university classes.

Abstract

A better understanding of redox (or electrochemical) processes in solution requires the knowledge of some basics related to chemistry. These include, what solvents and electrolytes are used in redox chemistry, the influence of parameters like conductivity, acidity, basicity, *pH*, and some thermodynamics and kinetics associated with these processes. This Section 1 summarizes some of these basics. Remember that all redox are chemical processes but not all chemical reactions are redox and only those involving the transfer of electrons are characterized as redox.

1. Aqueous and non-aqueous solvents

Aqueous (*aq*) solutions are based on water as the solvent. Substances can either be hydrophilic (polar) or hydrophobic (nonpolar) in contact with aqueous solutions[1-2]. Hydrophilic substances are attracted to water molecules to form soluble solutions, resulting in homogeneous mixtures. By contrast, hydrophobic substances have no affinity for water and tend to move away from water molecules, yielding separate phases. For example, NaCl (table salt) is a hydrophilic substance as it can fully dissolve in water to form an ionic solution containing weak complexes of Na^+ and Cl^- with H_2O molecules. By contrast, vegetable oil is hydrophobic as it tends to separate from water to form two distinct phases: water on the bottom and oil on the top. The reason behind solubility of substances (or solutes) in water is linked to the molecular forces exerted between water molecules and those of the solute species[3]. A solute dissolves in water if the attractive forces between its species and water molecules are stronger than those exerted between pure water molecules.

Non-aqueous solvents are any solvents other than water, which could be inorganic or organic in nature[5]. Organic solvents include acetone, alcohols, dimethyl sulfoxide, and tetrahydrofuran, among others. Inorganic solvents include sulfuric acid, phosphoric acid, sulfur dioxide, and hydrogen fluoride. A number of solutes are soluble in one or several of these non-aqueous solvents with variable degrees of solubility. Both aqueous and non-aqueous solvents are used in redox chemistry for either research studies or applications like in energy devices.

2. Electrolyte solutions

Solutes can be dissolved in aqueous and/or non-aqueous solvents to form electrolyte solutions[6-8]. Some solutes (e.g., NaCl) could fully dissolve in solvents like water to form strong electrolytes, able of conducting electricity. Other substances (e.g., some acids and bases) can partially dissolve in water to form weak electrolytes, often at solubility equilibrium states, where the dissolved ions are in equilibrium with the remaining insoluble solute. Some other solutes (e.g., glucose, lactose, and urea) can dissolve in solvents like water but without forming ion species. This category of solute solutions is called nonelectrolytes. Note that weak electrolytes weakly conduct electricity while non-electrolytes do not conduct electric current at all. These notions of electrolyte and ionic conductivity are very important in redox chemistry because the electric current in solution is transported by changed ionic species.

The mixture of two or more pure materials yields either a homogenous or heterogeneous solution. The species (molecules or ions) present in the solution can freely move around and collide with each other, increasing the probability of chemical reactions[9]. The mixture of different substances and possible occurring reactions are often expressed by molecular reactions (or equations). For example, the mixture of HCl with water yields an acid solution, expressed by the solubility of the acid in water: $HCl + H_2O \rightarrow H_3O^+ + Cl^-$. If a base like NaOH is added to this acidic solution, the base NaOH will firstly dissolve in water to form ionic species (Na^+, OH^-), then collision between H^+ and OH^- will form water molecules: $(H^+, Cl^-) + (Na^+, OH^-) \rightarrow (Na^+, Cl^-) + H_2O$. Note that the ions Na^+ and Cl^- did not actually participate in the reaction, thus called spectator ions. However, in other reactions (e.g., precipitation and complexation), these ions could participate in the precipitate or complex formation (e.g., $Ag^+ + Cl^- \rightarrow AgCl$). In this case, these ions are no longer spectators. Also, redox chemistry can occur in homogenous or heterogeneous mixtures, depending on the goal and applicability.

Chemical reactions might occur in one direction (forward) or both directions (forward and backward) until reaching equilibrium[10]. The corresponding reactions (or equations) are often presented by arrows symbolizing the type of the reaction that could be written as: $\rightarrow$ for forward reactions, $\Leftrightarrow$ for reactions occurring in both directions, and $\leftrightarrow$ for reactions at equilibrium states. For example, NaCl fully dissolves in water. Hence the reaction is presented by a forward arrow ($NaCl \rightarrow Na^+ + Cl^-$). However, ethanol can partially dissolve in water, thus presented by a double arrow ($C_2H_5OH \Leftrightarrow C_2H_5O^- + H^+$). When the equilibrium state of solubility is reached, the reaction could be written as: $C_2H_5OH \leftrightarrow C_2H_5O^- + H^+$. Keep in mind that redox processes could occur forward or in both directions until reaching equilibrium, but most redox calculated quantities are performed at equilibrium states.

3. Quantifying concentration of solutes

The amount of solute dissolved in a solvent can be quantified by calculating its molarity, normality, molality, or in terms of mole fraction, percent composition, and part per million[10]. Because individual atoms and molecules are invisible and hard to manipulate experimentally by atomic mass units, the notion of "mole" was put in place to facilitate the measurement of substances in the laboratory. A mole of any substance contains the Avogadro's number ($6.02214179 \times 10^{23}$) of its constituting particles, which might be atoms, molecules, ions, or species in general[11-12]. The molar mass of each known element representing the weight of 1 mole is often provided in the periodic table of elements. For substances composed of several atoms, the total mass is calculated by summing the molar masses of all the atoms.

The molarity (M) of a substance dissolved in a solvent is defined as the number of moles of solute present in a given volume (V) of the solution in litters: $M = \frac{number\ of\ moles\ (solute)}{V\ (solution)}$

The normality (N) is considered as the number of solute equivalents per liter (L) of solution: $N = \frac{number\ of\ equivalent\ solute}{L\ of\ solution}$. The term equivalent is used to take into account the stoichiometry of the involved species in the reaction.

The molality (m) expresses the solute concentration present in a given mass of the solvent in Kg: $m = \frac{number\ of\ moles\ (solute)}{Kg\ (solvent)}$

The mole fraction (x_i) represents the proportion of a certain component i relative to the sum of all components present in the solution: $x_i = \frac{mole\ fraction\ of\ the\ component}{total\ moles\ of\ all\ components}$. The sum of mole fractions for all components should give 1.

The percent composition expresses the relative content of each element in 100 g of the compound. Experimentally, the chemical composition of a compound could be identified using elemental analysis.

Parts per million (ppm) is often utilized for very small amounts of solutes dissolved in solvents. $ppm = \left(\frac{1\ part\ of\ substance}{1\ 000\ 000\ parts\ of\ solution}\right) \times 100\%$. For much smaller concentrations, parts per billion (10^9) and parts per trillion (10^{12}) are utilized to express the smaller levels of substances present in solutions.

Note that these notions are also used to express concentrations of redox species involved in redox processes.

4. **Conductivity of electrolyte solutions**

When a solute or electrolyte dissolves in a solvent like water to form charged anions and cations, the resulting solution becomes a conductor of electricity[8,13-14]. This type of electric conductivity is called ionic conductivity, which is different from the traditional conductivity occurring in metal conductors. The electric conductivity of a metal conductor is ensured by the formation of electrons (charged negatively) and holes (charged positively) through the metal's crystal structure. In electrolyte solutions, the ionic conductivity is ensured by the formation of a flow of charge upon the passage of an electric field. In other words, when the electrolyte solution is subjected to an electric current, with positive and negative polarities at both ends, the positively charged cations move towards the negatively charged pole and the negatively charged anions move to the opposite positive pole. This creates some sort of movement of charge that leads to electric conduction.

The basic laws used in conventional electricity, such as the Ohm's Law ($V = IR$, where V is the voltage, I is current, and R is the resistance of the solution) also apply to ionic conductivity of electrolyte solutions. In general, the ionic conductivity k (S m^2 mol^{-1}) is defined as: $k = \left(\frac{L}{RA}\right)$, where A is the cross sectional area of the electrodes, L is the distance between the two electrodes (negative and positive poles), and R is the solution resistance. The constants L and A are often determined from calibration with a cell with known conductivity. Experimentally, the ionic conductivities of solutions are determined using conductivity meters. Overall, the ionic conductivity of a solution increases as the concentration rises because more ions are present to induce more flow of charge. Note that the ionic conductivities of electrolytes solutions play a key role in redox chemistry[17].

5. Thermodynamics of electrolyte solutions

5.1. Energy quantities

A number of energy linked quantities applies to electrolyte solutions and eventually to redox processes, including internal energy, enthalpy, entropy, Gibbs free energy, chemical potential, and activity[16-18].

The internal energy (U) of an electrolyte solution could be seen as the sum of the total kinetic energies of all involved species and the potential energy induced by interactions between these species. Note that the kinetic energy results mainly from the motion of the species and potential energy associated with the attraction and repulsion between these species.

The change in enthalpy (*H*) of an electrolyte solution is associated with the variation in energy upon the addition or removal of heat at constant pressure. When a solution receives heat from an external source, the thermal energy is employed to overcome the internal bonding, and the reverse process occurs upon cooling to form bonding that holds the species together. A reaction is considered endothermic or consumes heat if *ΔH>0*, and exothermic or releases heat if ΔH<0.

The entropy (S) of an electrolyte solution can be viewed as a measure of the disorder of the species induced by motion, rotation, and vibration, among others. The more the electrolyte solution is complex in composition and reaction, the more the entropy is high because more energetic and spatial states are present within the system. The increase in temperature and concentration of soluble species raises the entropy due to the increase in motion of the species. Virtually, the entropy of any substance at *zero* Kelvin (0 K) is *zero*, meaning that at this temperature all the species are frozen in defined locations and the disorder becomes *zero*. The change in entropy during chemical reactions is calculated by subtracting the entropy values of the products from those of the reactants, taking into account the stoichiometric coefficients of both the reactants and products. For example, the entropy of dissociation of Na_2SO_4 ($Na_2SO_4 \rightarrow 2Na^+ + SO_2^{2-}$) is: $\Delta S = 2S(Na^+) + S(SO_2^{2-}) - S(Na_2SO_4)$.

Although the previous functions could occasionally appear in redox chemistry for demonstrative purposes, the most important thermodynamic quantity often utilized in electrochemistry is the Gibbs free energy (*G*). Spontaneous reactions always undergo a decline in enthalpy and an increase in entropy. However, processes subjected to reverses charges do not occur spontaneously and require external energy to achieve the transformation. Numerically, the change in Gibbs free energy at temperature *T* is defined by: *ΔG = ΔH - TΔS*, where *H* is the enthalpy, *S* is the entropy, and *T* is the temperature. A reaction is considered as spontaneous if *ΔG<0* and nonspontaneous if *ΔG>0*. At equilibrium *ΔG=0*, where both forward and backward processes occur at the same speed (or rate).

Chemical potential (μ) is another important quantity often employed in redox chemistry to express the change in the Gibbs free energy of a species at other constant parameters, such as pressure, temperature, and concentrations of the other species. Therefore, chemical potential is often called the partial molar free energy, and expressed as: $dG = -SdT + VdP + (\mu_1 dN_1 + \mu_2 dN_2 ...)$, where *dG* represents the infinitesimal change in Gibbs free energy, S is the entropy, V is the volume, *dT* and *dP* are the infinitesimal changes in temperature and pressure of the system, and *dNi* is the infinitesimal change in the number of species. At constant *P* and *T*, both infinitesimal changes of these variables are zero, leading to: $dG = (\mu_1 dN_1 + \mu_2 dN_2 ...)$. Since at equilibrium *dG*=0, hence all species at equilibrium have equal chemical potentials.

The activity of a species *i*, (a_i) is also a very important parameter frequently used in redox chemistry. The activity expresses the effective or real concentration of the species in a mixture because only diluted mixtures behave as ideal. The activity is directly related to the chemical potential: $\mu_i = \mu_i^o + RT \ln a_i$, where R is the gas constant, T is the temperature, and μ_i^0 is the chemical potential of the species *i* at the standard conditions. The concentration is often symbolized by () and concentration by []. For simplification reasons, all the formulas given in this manuscript are written as activities (), which in most cases represent the concentrations since most of the treated solutions are at the diluted states and behave as ideal (activity = concentration or () = []). By convention, the activities (or concentrations) of solids, pure substances, and substances in excess always equal to 1.

5.2. Chemical equilibria

In chemical reactions, reactants are consumed to produce products, moving the reactions forward from left to right. However, because most chemical reactions do not reach completion, the transformation of the reactants into products stops at a certain point, letting both the forward and reverse reactions to proceed at the same speed (or rate). This state is determined as the equilibrium[16-18]. The equilibrium could either be homogenous occurring in the same phase or heterogeneous proceeding in differences phases. Note that redox equilibria could be homogenous or heterogeneous, depending on whether they occur in one or several phases.

The disturbance of a chemical equilibrium through changes in concentrations of reactants or products, temperature or pressure should push the reaction to move in one particular direction to restore the equilibrium state (Le Chatelier's principle).

At equilibrium, the ratio of the reactants and products is governed by the law of mass action and quantified by calculating the equilibrium constant (K_{eq}), expressed by the ratio of the activities (or concentrations) of the products to those of the reactants by considering their stoichiometry coefficients. For example, the equilibrium constant of the dissolution reaction of H_2SO_4 in water ($H_2SO_4 \leftrightarrow 2H^+ + SO_4^{2-}$) is: $K_{eq} = \frac{(H^+)^2(SO_4^{2-})}{(H_2SO_4)}$

Note that the reaction could move in both directions depending on the condition. For example, the addition of SO_4^{2-} or H^+ should move the equilibrium to the left (Le Chatelier's principle) but more H_2SO_4 should move the reaction to the right to consume the excess H_2SO_4 and restore the equilibrium state. The knowledge of concentrations will allow determining the equilibrium constant, and vice versa the knowledge of the equilibrium constant will allow estimating the concentrations if the stoichiometric coefficients of the reactants and products are known. The equilibrium constant is linked to the Gibbs free energy by the expression: $\Delta G = -RT \ln K_{eq}$. Note that K_{eq} is affected by the temperature of the system.

5.3. Some important reaction equilibria

Some reactions have very well-known equilibrium constants, including solubility, precipitation, complexation, acid-base, and redox reactions[16-18]. The focus of this manuscript is the redox equilibria but solubility, precipitation, complexation and acid-base equilibria could sometimes be involved in redox processes, thus they are briefly reminded. In all cases, chemical reactions, including redox processes, always have to be mass and charge balanced, as they are written to describe the transformation of a certain number of moles of reactants into products. In other words, chemical reactions are always governed by the law of mass conservation, meaning that atoms during the transformation remain the same, and cannot be created or destroyed in the overall process. Therefore, the number of atoms on both sides of the reaction must remain the same. The number of moles of each element involved in the reaction is indicated by its stoichiometric coefficient. The total number of atoms of each element is obtained by first multiplying the stoichiometric coefficient by the subscript then followed by summing all atoms of each element on each side of the reaction. At the end, the mass and charge on both sides of the reaction must be verified and equalized if necessary.

5.4. Solubility, precipitation, and complexation reactions

The mixture of an ionic solute (or electrolyte) with a solvent like water should dissociate the solute into its anionic species (cations and anions). Hence, solubility could be expressed in terms of how much mass of the solid solute is dissociated in a given volume of the solvent. The solubility depends on whether the forces holding the crystal structure of the solute could be broken in contact with the solvent molecules. By contrast, the precipitation and complexation reactions could be regarded as the reverse reactions of solubility. In precipitation and complexation, dissolved ions combine together to form a precipitate or complex, depending on the conditions. A precipitate is often neutral while a complex is charged. For example, AgCl formed by the combination of Ag^+ with Cl^- ions is a precipitate, while $[Fe(CN)_6]^{3-}$ induced by the combination between Fe^{2+} with six CN^- is a complex.

At equilibrium, the remaining undissociated solid solute is in contact with the solution saturated with dissolved ions. At this stage, both processes, solubility and precipitation or complexation (depending on the species and conditions), occur at equal rates (forward and reverse). Overall, if we consider an ionic solute (A_nB_m) dissociating in water into its ionic species: $A_nB_m \Leftrightarrow nA^{m+} + mB^{n-}$, the solubility constant (K_s) is written as:

$$K_s = \frac{(A^{m+})^n (B^{n-})^m}{(A_nB_m)}$$

In this case, the precipitation or complexation, could be presented by the reverse reaction: $nA^{m+} + mB^{n-} \Leftrightarrow A_nB_m$, with the precipitation constant (K_p) or complexation constant (K_c) written as the reverse of K_s: K_p (or K_p) $= \frac{1}{K_s}$.

5.5. Acid and base reactions

Most acid and base substances are electrolytes. When put in contact with a solvent like water, acids and bases dissociate to form cation and anion species. Throughout history, acids and bases have been expressed by many definitions, including Arrhenius, Lewis, and Brønsted/Lowry. The Arrhenius definition suggests that acids liberate H^+ ions in water and bases form OH^- ions. Thus, the net reaction between an acid and a base produces water molecule ($H^+ + OH^- \rightarrow H_2O$). The Lewis definition proposes that bases give away unshared electron pairs to acids, and the Brønsted/Lowry suggests that acids transfer a proton to bases. The relevance of each definition often depends on the case study. For example, in redox chemistry, the Lewis definition is sometimes more useful to explain electron transfer from one species to another. However, overall, the Brønsted/Lowry theory is the most employed concept to explain acid and base processes through the transfer of a proton from the acid to the base.

The dissociation degree of acid and base substances into ionic species depends on the substance structure and the forces holding its structural network. Weak acids and bases will moderately dissociate in the solvent while strong acids and bases will fully dissociate. For example, HCl is a strong acid, thus it will fully dissociate in water. However, methanol is a weak acid and will partially dissociate in the same solvent. Note that water molecule could play either the role of a weak acid or weak base, depending on the reaction and conditions: $2H_2O_{(l)} \leftrightarrow$

$H_3O^+_{(aq)}$ + $OH^-_{(aq)}$. This water dissociation reaction is characterized by an equilibrium constant, called the ion-product constant of water (K_w): $K_w = \frac{(H_3O^+)(OH^-)}{(H_2O)^2}$. Because water is present in excess, its concentration is 1, thus $K_w = (H_3O^+)(OH^-) = 10^{-14}$ at 25° C, and $(H^+) = (OH^-) = 10^{-7}$ mol L^{-1} for pure water. The equilibrium constant of any acid and base reaction, often called K_a for acids and K_b for bases, is determined from the dissociation reaction. The product of K_a with K_b always gives K_w according to: $(K_a)(K_b) = K_w$

Because the values of equilibrium constants are often very small, *pK* obtained by the negative base-10 logarithm of the ionic molarity, is put in place to make more sense of these quantities and facilitate comparison between the properties of species. Most equilibrium constants, including K_s, K_p, K_c, K_w, K_a and K_b are expressed in terms of *pK* (*pK$_s$*, *pK$_p$*, *pK$_c$*, *pK$_w$*, *pK$_a$*, and *pK$_b$*). Also, the concentration of protons ($H^+_{(aq)}$), reflecting the acidity of a solution is often expressed by pH: $pH = -Log(H^+) = Log \frac{1}{(H^+)}$, used as a measurement scale of acidity and basicity of solutions. For pure water, $(H^+) = 10^{-7}$ M, and hence pH = 7. Acidic and basic solutions have respectively: $(H^+) > 10^{-7}$ and $< 10^{-7}$ M, thus respective pH < 7 and > 7. Similarly, as the ion-product constant of water is: $K_w = (H_3O^+)(OH^-) = 10^{-14}$ at 25° C, $pK_w = pH + pOH = 14$.

Some media, such as those used for biological or biochemical culture, require constant amounts of acid and base species because drastic charges of acidity could result in denaturation or death of the biological species. Buffer solutions with high resistance to changes in pH are thus required for biological, biochemical, and similar media. These buffers solutions contain conjugate acid-base pairs ensuring to maintain the acid and base equilibria when small amounts of acids or bases are added to the buffer. Buffers are prepared by mixing weak acid or base with its salts. Typical employed buffers conjugates are based on phosphates, acetates, and citrates, among others. For example, the acid/base conjugate used in phosphate buffer solutions is: HPO_4^{2-}/PO_4^{3-}. Thus, the addition of small amounts of H^+ should move the equilibrium in the direction that will consume the excess protons (Le Chatelier's principle), and vice versa the addition of a base should move the equilibrium in the reverse direction that will consume it and restore the original pH. However, each buffer represents its limitation when it comes to how much acid or base could be neutralized before large changes in pH appear. This is called the buffer capacity, which differs from one buffer to another. Note that natural media like blood and seawater have several acid/base conjugates allowing them to maintain constant pH values.

For analytical purposes, neutralization of an acid with a base or vice versa could sometimes be helpful in determining unknown concentrations or even species. This is called acid/base neutralization reactions, which leads to the formation of the water molecules and salt. The progress of the titration reaction could be followed by means of indicators (e.g, phenolphthalein) or pH meters. In neutralization reactions, acid or base with the unknown concentration (or titrant) is added to an acid or base with known concentration. A curve showing the variation of pH as a function of the added volume of the solution with unknown concentration could thus be drawn. At the endpoint (or equivalent point), the number of equivalents (or normality) of both the acid and base are identical. This state is characterized by a sudden jump in pH on the titration curve. At this point: $C_{unknown}V_{unknown} = C_{known}(V_{final} - V_{initial})$, where $V_{initial}$ represents the initial volume, V_{final} is the final volume at the equivalent point, C_{known} is the concentration of the known solution, and $V_{unknown}$ is the volume of the unknown solution. This relationship should allow calculating the unknown concentration since the other parameters are all known. Note that the concentration should be expressed in terms of normality to take into account the number of the equivalent of the transferred protons.

It is important to remember that most redox reactions are affected by the acidity of the solution. Therefore, a better understanding of acid and base reactions will greatly help in the understanding the electrochemical processes.

6. Redox reactions

This will be discussed in other sections.

7. Kinetics of chemical reactions

During chemical processes, the reactants collide with each other to form the products and the speed of this conversion defines the kinetics of the reaction[19-20]. To form the products, the collision between the reactant species requires a minimum energy to overcome the barrier of products formation. This is called the activation energy (E_a). Note that not all collisions lead to the formation of products because not all occur with either sufficient energy or proper orientation of the reactant species during the collision process. The initial stage of this transformation process is known as the transition state, characterized by the breakdown of the old bonds to form new ones at the intermediate states, known as the activated complex.

Sometimes the collision between the reactant species is so slow that the reaction will not reach completion even after extended periods of time. To accelerate the course of the reaction under these circumstances, influencing parameters like temperature, concentration, and use of catalyst materials could be manipulated to reach this goal. The rate of most elementary reactions increases as the temperature rises due to the increased in the kinetic energy of the colliding species, boosting the probability of collisions to reach the activation energy. Also, raising the concentration of the reactants should force more species to collide, which would raise the kinetics. A catalyst material is a species that could temporarily combine with the reactant to decrease its activation energy and allow the process to speed up. Once the reaction reached completion, the catalyst will be released to the reacting media and could be separated from the products and the remaining reactants. Catalysts could be used homogenously or heterogeneously. In the former case, the catalyst is mixed with the reactants to interact with them whereas in the latter case, the catalyst is immobilized in a different phase (e.g., solid) and the reactant species must travel to the surface covered with the catalyst to be converted into products.

The change in concentration of the reactants or products as a function of time defines the reaction rate: $rate = \frac{dC}{dt}$, where C is the concentration of either the reactant or product since what is consumed must equal what is produced, and t is the necessary time spent in inducing the change in concentration. The knowledge of the elementary steps with their rates should allow determining the reaction mechanism from the start until completion of the process. Most reaction mechanisms are complex and involve multi-step processes with the formation of intermediates. Examples of these mechanisms include first order and second order kinetics, expressed as: *rate* = *k*(A) for first order and *rate* = *k*(A)^2 or *rate* = *k*(A)(B) for second order kinetics. A and B represent the concentrations (or activities) of the reactants.

Although the kinetics of redox processes could also be evaluated, for simplicity reasons, most processes are examined at equilibrium.

8. Summary of important chemistry units used in electrochemistry

Table 1: Summary of some important chemistry units used in electrochemistry.

Quantity	Symbol	Unit	Other units
Volume	V	cubic meter	m^3

		liter	$dm^3 = 10^{-3}\ m^3$
Mass	u or Da	unified atomic mass unit	$(6.022 \times 10^{23})^{-1}\ g$
	m	gram, kilogram	$g = 10^{-3}\ Kg$
Time	s, min, h, d	second, minute, hour, day	1 day = 24 h = 1440 min = 86400 s
Area	A	square meter	m^2
Temperature	°C	degrees Celsius	K-273.15°
Molarity	M	mol L^{-1}	10^{-3} mmol L^{-1}
Molality	*m*	mol Kg^{-1}	10^{-3} mmol Kg^{-1}
Normality	N	eq L^{-1}	
Reaction rate		mol L^{-1} s^{-1}	10^{-3} mmol L^{-1} s^{-1}
Pressure	P	pascal	$N\ m^{-2} = kg\ m^{-1}\ s^{-2}$
	atm	atmosphere	101325 Pa
Energy	J	joule	$N\ m = m^3\ Pa = m^2\ kg\ s^{-2}$
	Cal	calorie	4184 J
Power	W	watt	J s^{-1}
Electric charge	C	coulomb	A s
Electric potential and electromotive force	V	volt	W A^{-1}

Summary

Chemical reactions, including redox processes, could be performed in aqueous or non-aqueous solvents, depending on the affinity between the solute species and the solvent. Water molecules are attracted to hydrophilic but repulsed by hydrophobic substances. The dissolution of electrolytes in solvents could either be full or partial, depending on the forces exerted between the solvent molecules and those of the electrolyte. The concentration of dissolved solutes in solvents could be expressed in many ways, including molarity, molality, normality, mole fraction, percent composition, and part per million. Fully dissociated electrolytes induce solutions with great ionic conductivities whereas partially dissociated electrolytes lead to weak ionic conductivities. The mixture of two or more substances induces homogenous or heterogeneous solutions. Chemical reactions, including redox processes, could occur in homogenous or heterogeneous mixtures to transform reactants into products. In the beginning of the transformation process, reactions proceed in the forward direction until reaching equilibrium, where both forward and reverse reactions occur at the same speed (or rate). The changes in energy of chemical reactions, including redox processes, can be described by several thermodynamic quantities, including enthalpy, entropy, Gibbs free energy, and chemical potential. Solubility, complexation, precipitation, acid/base, and redox processes are the most well-studied processes. These reactions often involve multiple steps and their identifications should help in determining complex reaction mechanisms. Electrochemical (or redox reactions) are among the most studied chemical reactions, and some of their basics are discussed in the next sections.

References

1. Silverstein, T. P. (1998), The Real Reason Why Oil and Water Don't Mix, Journal of Chemical Education, 75: 116-346.
2. Wenzel, R. N. (1936), Resistance of Solid Surfaces to Wetting by Water, Industrial and Engineering Chemistry, 28 (8): 988-994.
3. Gong, Y; Grant, Brittain, H. G. (2007), Solvent Systems and Their Selection in Pharmaceutics and Biopharmaceutics, Principles of Solubility, pp 1-27, Part of the Biotechnology: Pharmaceutical Aspects book series (PHARMASP, volume VI). Springer.
4. Covington, A. (2012), Physical Chemistry of Organic Solvent Systems. Springer Science & Business Media.
5. Chipperfield, J. (1999), Non-Aqueous Solvents, Oxford University Press.
6. Ronald Fawcett, W. (2004), Liquids, Solutions, and Interfaces, Oxford University Press.
7. Robinson, R. A; and Robert Harold Stokes, R. S. (2002), Electrolyte solutions. Courier Corporation.
8. Wright, M. R. (2007). An Introduction to Aqueous Electrolyte Solutions, Wiley.

9. Keeler, J.; Wothers, P. (2003), Why Chemical Reactions Happen, Oxford University Press.

10. IUPAC. Compendium of Chemical Terminology, 2nd ed. (the "Gold Book"). Compiled by McNaught, A. D.; Wilkinson, A. (1997), Blackwell Scientific Publications, Oxford.

11. Avogadro, A. (1811), Essai d'une Maniere de Determiner les Masses Relatives des Molecules Elementaires des Corps, et les Proportions selon Lesquelles elles Entrent dans ces Combinaisons, Journal de Physique, 73: 58-76.

12. De Bieve, P; Peiser, H. S. (1992)., Atomic Weight: The Name, Its History, Definition and Units". Pure and Applied Chemistry. 64 (10): 1535-43.

13. Gray, J. R. (2004), Conductivity Analyzers and Their Application, In Down, R. D.; Lehr, J. H. Environmental Instrumentation and Analysis Handbook. Wiley. pp. 491–510.

14. Marija, B. R; Dušan H. (2006), Modern Advances in Electrical Conductivity Measurements of Solutions, Acta Chimica Slovenica, 53, 391-395.

15. Bockris, J. O. M.; Reddy, A. K. N; Gamboa-Aldeco, M. (1998), Modern Electrochemistry (2nd. ed.). Springer.

16. Tschoegl, N. W. (2000), Fundamentals of Equilibrium and Steady-State Thermodynamics, Elsevier, Amsterdam.

17. Atkins, P.; Julio D. P. (2006), Physical Chemistry, 8th ed. Oxford University Press.

18. Salzman, W. R. (2001), Open Systems, Chemical Thermodynamics. University of Arizona.

19. Guy M.; Gregory S. Y. (2011), Kinetics of Chemical Reactions, Wiley.

20. Jorge A. (2017), Chemical Reaction Kinetics: Concepts, Methods and Case Studies, Wiley.

Practical Questions and Problems with Solutions

A set of practical questions and problems with detailed solutions are provided to better understand the discussed concepts. The questions and problems range from simple to complex.

Q1. i) Provide few examples of aqueous and non-aqueous solvents. ii) Propose a method for preparing electrolyte solutions using these solvents. iii) Which types of solvents are used in electrochemical studies?

Ans1. i) Aqueous solvents mean water, and non-aqueous solvents could either be organic (e.g., acetone, acetonitrile) or inorganic (e.g., HCl, H_2SO_4). ii) Electrolyte solutions are prepared by adding solutes into solvents. For example, the addition of KCl into water forms an aqueous electrolyte while the addition of NH_4Cl into organic solvents forms non-aqueous electrolytes. iii) Both types of solvents are used in electrochemistry, depending on the solubility of the redox species and the aim of the study.

Q2. i) How many molecules of water, oxygen, and hydrogen are present in 1 mole? iii) Does this differ from the number of atoms of gold and iron present in 1 mole?

Ans2. i) 1 mole of water, oxygen or hydrogen contains exactly $6.02214179 \times 10^{23}$ molecules (Avogadro's number). ii) The same applies to pure metals (e.g., gold (Au) or iron (Fe)), where 1 mole of each element contains $6.02214179 \times 10^{23}$ atoms.

Q3. A small quantity of pure iron (Fe) is collected at nearby mine and was taken to a local chemistry lab for analysis. i) In your view what equipment is required to determine the number of moles present in the bag of iron? ii) If the sample weighs 5.1 Kg, how many atoms of Fe are present in the bag?

Ans3. i) The chemist requires a balance to weight the iron and determine its mass.

ii) If the mass of iron is 5.1 Kg, this needs to be first converted into the number of moles. From the periodic table, 1 mole of Fe weights 55.8 g. Thus, 5.1×103 g contains: $\frac{5.1 \times 10^3}{55.8} =$ 91.4 moles of Fe. Since 1 mole of Fe contains the Avogadro's number or $6.02214179 \times 10^{23}$ atoms of Fe, the total number of moles in the bag is: $91.4 \times 6.02214179 \times 10^{23} = 5.5 \times 10^{25}$ atoms of Fe.

Q4. i) How many grams exist in 1 mole of Au, S, H_2S, and NH_3? ii) One mole of these substances contains how many atoms or molecules?

An4. i) The periodic table provides numerous information, such as the molar mass of each element. The atomic masses of Au, S, H and N are 196.96, 32.06, 1.00 and 14.00 g mol^{-1}, respectively. For molecules, the molecular weight is determined by adding the molar masses of all atoms present in the molecule. For example, mass (H_2S) = mass (H) × 2 + mass (S) = (1.00 ×

2) + 32.06 = 34.06 g mol^{-1}. Similarly, mass (NH$_3$) = mass (N) + mass (H) × 3 = 14.00 + 1.00 × 3 = 17 g mol^{-1}.

ii) 1 mole of each of substance contains the same number of atoms or molecules, which is the Avogadro's number or 6.02214179 × 10^{23}.

Q5. i) Define an equilibrium state of a chemical reaction (A$^+$ + B$^-$ → AB) at constant temperature and pressure. ii) What is the solubility constant of the reaction: AB → A$^+$ + B$^-$? iii) Arrange the following salts by increasing solubility: Cu(OH)$_2$ (K_s = 2.2 × 10^{-20}), Ca(OH)$_2$ (K_s = 8.0 × 10^{-6}), and Al(OH)$_3$ (K_s = 1.8 × 10^{-33}).

Ans5. i) A chemical reaction (A$^+$ + B$^-$ → AB) reached equilibrium when the conversion rate of the reactants (A$^+$ and B$^-$) into the product (AB) is equal to the conversion rate of the reverse reaction (AB into A$^+$ and B$^-$). In other words, both forward and reverse reactions occur at the same rate.

ii) The solubility product of a compound is defined as the product of the molar activities (or concentrations) of the ions raised to the power of their respective stoichiometric coefficients in the equilibrium reaction. For the given reaction: $K_s = \frac{(A^+)(B^-)}{(AB)}$. Activity is used for real solutions while concentration is employed for diluted solutions as they behave as ideal. If AB is a solid, its activity (or concentration) is by convention 1. Thus, $K_s = (A^+)(B^-)$.

These salts could be compared according to their K_s values. The more the solubility constant is high, the more the concentration of the species in solution is elevated. Therefore, solubility decreases in the following order: Al(OH)$_3$ < Cu(OH)$_2$ < Ca(OH)$_2$.

Q6. Consider an aqueous solution of metal ions containing Ag$^+$ (0.01M) and Hg$_2^{2+}$ (0.01M). i) In your view could these ions precipitate if they are in contact with halogen ions, such as Cl$^-$ or I$^-$? ii) Which of the two salts (AgI or Hg$_2$I$_2$) should precipitate first and why? The solubility constants K_s (AgI) = 8.5 × 10^{-17} and K_s (Hg$_2$I$_2$) = 2.5 × 10^{-26}.

Ans6. i) Yes, if Ag^+ or Hg^{2+} is in contact with Cl^- or I^-, slightly soluble salts should precipitate to form AgCl, AgI, HgCl$_2$, and HgI$_2$. ii) The amount of the precipitated salt depends on the concentration of the ion species present in the solution.

The precipitation reaction with I^- ions could be summarized as follows:

$Ag^+ + I^- \rightarrow AgI$, with $K_p = \dfrac{(AgI)}{(Ag^+)(Cl^-)} = \dfrac{1}{K_s}$

Note that (AgI) = 1 because it is solid.

The concentration of I^- could be calculated from the given data: $8.5 \times 10^{-17} = (0.01)(I^-)$. Hence, the concentration of I^- allowing AgI to precipitate is 8.5×10^{-15} mol L^{-1}.

For $2Hg^+ + 2I^- \rightarrow Hg_2I_2$, $K_p = \dfrac{(Hg_2I_2)}{(Ag^+)^2(I^-)^2} = \dfrac{1}{K_s}$

Also, $(Hg_2I_2) = 1$ because it is solid.

Therefore, the concentration of I^- could be calculated from the relationship: $2.5 \times 10^{-26} = (0.01)^2 (I^-)^2$. Hence, the concentration allowing Hg$_2$I$_2$ to precipitate is 1.58×10^{-12} mol L^{-1}.

It can be seen that the estimated concentration of (I^-) precipitating AgI is lower, meaning that AgI requires less ionic concentration in solution to precipitate. As a result, AgI will begin precipitating first and as I^- is added to the solution, Hg$_2$I$_2$ will progressively form.

Q7. Calcium sulfate (CaSO$_4$) is a slightly soluble salt. A solution of CaSO$_4$ is prepared by dissolving 0.56 g L^{-1} in water. Estimate the solubility constant of this salt at 25°C. The molar mass of CaSO$_4$ = 136.2 g mol^{-1}

Ans7. The solubility equation of CaSO$_4$ in water could be written as follows:

$CaSO_{4(s)} \rightarrow Ca^{2+}{}_{(aq)} + SO_4^{2-}{}_{(aq)}$, with $K_p = \dfrac{(Ca^{2+})(SO_4^{2-})}{(CaSO_4)} = \dfrac{1}{K_s}$

Note that the concentration of CaSO$_{4(s)}$ is 1 because it is solid.

The reaction stoichiometry indicates that 1 mole of CaSO$_4$(s) gives 1 mole of Ca^{2+} and 1 mole of SO_4^{2-}. Since the concentration is given in g L^{-1}, thus it should be converted to mol L^{-1}.

$(CaSO_4) = \dfrac{0.56}{136.2} = 4.1 \times 10^{-3} = (Ca^{2+}) = (SO_4^{2-})$

This gives: $K_s = (Ca^{2+})(SO_4^{2-}) = (4.1 \times 10^{-3})^2 = 1.68 \times 10^{-5}$

Q8. The reaction between Cu^{2+} and OH^- induces a slightly soluble metal hydroxide, Cu(OH)$_2$. i) Write down the reaction of formation of Cu(OH)$_2$. ii) How many moles of OH^- and Cu^{2+} are required to form 1 mole Cu(OH)$_2$? iii) If the solubility constant of Cu(OH)$_2$ is 2.2×10^{-20}, estimate the concentrations of the soluble ions in solution.

Ans8. i) The formation reaction can be written as follows: $Cu^{2+} + 2OH^- \rightarrow Cu(OH)_2$

ii) The reaction indicates that 1 mole Cu^{2+} requires 2 moles of OH^- to form 1 mole $Cu(OH)_2$.

iii) The solubility constant of $Cu(OH)_2$ is the reverse value of the precipitation constant.

$K_s = (Cu^{2+})(OH^-)^2$, with $= (Cu^{2+}) = (OH^-)$

Assuming that $(Cu^{2+}) = C$, hence $(OH^-) = 2C$ and $K_s = (C)(2C)^2 = 4C^3$.

The replacement of K_s by its value gives: $C^3 = \frac{2.2 \times 10^{-20}}{4}$, or $C = 1.8 \times 10^{-7}$ mol L^{-1}

Q9. H_3PO_4 reacts with NaOH to produce NaH_2PO_4. i) In what category should you place this reaction? What are the roles of H_3PO_4 and NaOH? ii) What is the number of equivalents of the protons involved during the first dissociation?

Ans9. i) The reaction of H_3PO_4 with NaOH is primarily an acid/base reaction, where H_3PO_4 plays the role of an acid and NaOH is the base.

ii) This reaction could be summarized as follows: $H_3PO_4 + NaOH \rightarrow NaH_2PO_4 + H_2O$

During this first dissociation, one equivalent H_3PO_4 per mole produces one equivalent of the protons H^+. However, if the reaction continues dissociating, up to 3 equivalents of protons could be produced.

Q10. i) What is the best way to prepare an aqueous acidic solution, basic solution, and neutral solution? 5.33 g H_2SO_4 are added to 48.00g water to form 100.00 mL sulfuric acid solution. ii) Estimate the molality, molarity, and normality of the solution. The molar mass of $H_2SO_4 = 98.07$ g mol^{-1}.

Ans10. i) The best way to prepare an acidic solution is by adding an acid to water. To prepare basic solution, a base should be added to water. For neutral solution, a salt (e.g., NaCl) can be added to water. The pH of each solution could be monitored by a pH meter (pH ~7 for neutral, pH<7 for acidic, and pH>7 for alkaline).

ii) The addition of sulfuric acid into water produces an aqueous acidic solution.

Molarity M is defined as the number of solute moles present in a given volume V of the solution:

$$M = \frac{\text{moles of solute}}{\text{volume of solution}} = \frac{5.33\,g}{(98.07\,gmol^{-1})\,(0.1\,L)} = 0.543 \text{ mol L}^{-1}$$

Molality m is expressed in terms of how much solute is present in a given mass of the solvent in Kg:

$$m = \frac{\text{moles of solute}}{Kg \text{ of solvent}} = \frac{5.33\,g}{(98.07\,gmol^{-1})\,(0.048\,Kg)} = 1.13 \text{ mol Kg}^{-1}$$

The normality N is considered as the number of solute equivalents per liter L of solution: $N = \frac{\text{equivalents of solute}}{L \text{ of solution}} = 2 \times M = 2 \times 0.543 = 1.086$ N

Q11. i) In what category should you place Ba(OH)$_2$ and HCl? 0.1 M Ba(OH)$_2$ is added to 0.08 M HCl and allowed to react for some time. ii) Estimate the molarity of the ions left after completion of the reaction.

Ans11. i) HCl is an acid and Ba(OH)$_2$ is a base. The mixture of both should produce water and salt according to the following acid/base neutralization reaction:

2HCl + Ba(OH)$_2$ → BaCl$_2$ + 2H$_2$O

ii) The stoichiometry of the reaction indicates that 1 mole Ba(OH)$_2$ requires 2 moles HCl. Thus, 0.08 moles HCl should require (0.5 × 0.08 = 0.04) moles of Ba(OH)$_2$. Because 0.1 moles of Ba(OH)$_2$ is used, hence (0.1 − 0.04) = 0.06 moles Ba(OH)$_2$ will remain after completion of reaction.

In sum, after completion of the reaction, all the HCl is consumed leaving 0.06 M unreacted Ba(OH)$_2$ along with water molecules and 0.04 M BaCl$_2$.

Q12. i) Provide few examples of monoacids, diacids, and triacids. ii) What is the difference between the three acid categories? iii) Estimate the number of grams of Mg(OH)$_2$ required to completely convert 40.0 ml of 0.206 N H$_3$PO$_4$ into PO$_4^{3-}$. The molar mass of H$_3$PO$_4$ = 98 g mol^{-1}

Ans12. i) Examples of monoacids, diacids, and triacids are HCl, H$_2$SO$_4$ and H$_3$PO$_4$, respectively.

ii) The difference between these acids deals with the number of protons that could be donated during dissociation. Monoacids can only donate one proton whereas diacids and triacids could donate up to two and three, respectively.

iii) The reaction between the base Mg(OH)$_2$ and triacid H$_3$PO$_4$ induces water and salt. The neutralization reaction could be summarized as follows:

3Mg(OH)$_2$ + 2H$_3$PO$_4$ → Mg$_3$(PO$_4$)$_2$ + 6H$_2$O

The mass of H$_3$PO$_4$ present in 40 ml of 0.206 N is: $\frac{(98 \text{ gmol}^{-1}) \times 0.206 \times 0.04}{3} = 0.269$ g (use definition of normality). The number of moles is estimated as: $\frac{0.269 \text{ g}}{98 \text{ gmol}^{-1}} = 0.0027$ moles. The reactions indicates that 0.0027 moles of H$_3$PO$_4$ requires $\frac{3}{2 \times 0.0027} = 0.00405$ moles of Mg(OH)$_2$. Hence, the mass corresponding to 0.00405 moles of Mg(OH)$_2$ = 58.32 × 0.00405 = 0.236 g

Q13. i) Describe what happened during an acid-base neutralization reaction. ii) Estimate the volume of HCl (0.200 M) required to neutralize 0.70 g of calcium hydroxide dissolved in water. The molar mass of Ca(OH)$_2$ is 74.09 g mol^{-1}.

Ans13. i) During the neutralization reaction between an acid and a base, the proton liberated by the acid will react with the hydroxyl generated by the base to form water molecules. The overall reaction should produce water molecules and salt ions.

The neutralization of CaCl$_2$ with NaOH could be summarized as follows:

2HCl + Ca(OH)$_2$ → CaCl$_2$ + 2H$_2$O

The stoichiometry of the reaction indicates that 1 mole Ca(OH)$_2$ requires 2 moles HCl for neutralization. The number of moles of Ca(OH)$_2$ is: $\frac{0.70\ g}{74.09\ gmol^{-1}}$ = 0.0094 moles. Thus, 0.0094 moles of Ca(OH)$_2$ should require 2 × 0.0094 = 0.0188 moles of HCl.

The molarity M is defined as the number of moles of solute present in a given volume V of the solution: $M = \frac{moles\ of\ solute}{volume\ of\ solution}$. Thus, $V = \frac{moles\ of\ solute}{M} = \frac{0.0188}{0.200} = 0.094\ L$

In sum, 94 ml of the HCl solution are required to neutralize the base.

Q14. Lysergic acid is an organic acid with the formula C$_{15}$H$_{15}$N$_2$COOH. i) Write down the dissociation reaction of this acid in water. ii) Lysergic acid is titrated by NaOH, identify the neutralization reaction. iii) Calculate the number of grams of C$_{15}$H$_{15}$N$_2$COOH required to neutralize 4.3 ml of 0.05 M NaOH solution to the equivalence point. The molar mass of C$_{15}$H$_{15}$N$_2$COOH = 268 g mol^{-1}.

Ans14. i) The dissociation of C$_{15}$H$_{15}$N$_2$COOH in water could be summarized by the following reaction:

C$_{15}$H$_{15}$N$_2$COOH → C$_{15}$H$_{15}$N$_2$COO$^-$ + H$^+$

ii) The neutralization reaction with NaOH induces a salt and water molecules according to the reaction:

C$_{15}$H$_{15}$N$_2$COOH + NaOH → C$_{15}$H$_{15}$N$_2$COONa + H$_2$O

iii) The stoichiometry of the reaction indicates that 1 mole C$_{15}$H$_{15}$N$_2$COOH requires 1 mole NaOH.

The number of moles of NaOH = 0.043 × 0.05 = 0.00215 moles. According to the reaction, 0.00215 moles of NaOH requires 0.00215 moles of lysergic acid.

The number of grams corresponding to 0.00215 moles lysergic acid = 0.00215 × 268 = 0.576 g

In sum, 0.576 grams of lysergic acid are required to drive the neutralization reaction with NaOH to the equivalent point.

Q15. i) Provide few examples of organic acids. ii) Can you determine the structure of any of these acids using neutralization reactions with known bases? 0.200 g of unknown acid dissolved in 100 mL of water is titrated to the endpoint with 22.6 ml of 0.200 M NaOH. iii) Determine whether the unknown acid is among the proposed three possibilities: CH_3COOH, $CH_3COCOOH$, and/or CH_3CH_2COOH. The molar masses of these acids are: CH_3COOH = 60 g mol^{-1}, $CH_3COCOOH$ = 88 g mol^{-1}, and CH_3CH_2COOH = 74 g mol^{-1}.

Ans15. Examples of organic acids are: CH_3COOH, $CH_3COCOOH$, and CH_3CH_2COOH (the proposed ones). ii) Yes, by knowing the bases and the neutralization conditions, it is possible to determine the structure of the acid.

iii) The reaction between each of the proposed acids with NaOH should induce a salt and water according to the reactions:

$CH_3COOH + NaOH \rightarrow CH_3COONa + H_2O$

$CH_3COCOOH + NaOH \rightarrow CH_3COCOONa + H_2O$

$CH_3CH_2COOH + NaOH \rightarrow CH_3CH_2COONa + H_2O$

The stoichiometries of the reactions indicate that 1 mole of each acid reacts with 1 mole of NaOH. The number of moles of NaOH reacting with each acid is 0.0226 × 0.200 = 0.00452 moles NaOH. According to the stoichiometry, this should also be the number of moles of each acid.

Therefore, the molar mass of the unknown acid is: $\frac{0.200}{0.00452 \times 0.1}$ = 442.47 g mol^{-1}

None of the provided molar masses matches the calculated molar mass. Thus, the used acid for neutralization is not among the listed ones.

Q16. NaOH is dissolved in water to form a 0.02 M NaOH solution. Similarly, acetic acid is dissolved in water to form a 0.14 M acetic acid solution. i) Estimate the pH of both the NaOH and acetic acid solutions. ii) If the acetic acid solution at 0.14 M is only ionized at 1.4%, what would be its pH?

Ans16. i) The dissolution of NaOH and acetic acid in water could be expressed by the following reactions:

$NaOH \rightarrow Na^+ + OH^-$

$CH_3COOH \rightarrow CH_3COO^- + H^+$

The concentration or molarity of the NaOH is 0.02M, and the stoichiometry of the reaction indicates that 1 mole NaOH gives 1 mole of OH⁻. Thus, the pOH of the solution is: $pOH = -\text{Log}(OH^-) = 1.7$, and because $pH + pOH = 14$, hence $pH = 12.3$.

Similarly, the molarity of the acetic acid solution is 0.14M, which will also be the concentration of the H^+ ions according to the stoichiometry of the reaction. pH is defined by the formula: $pH = -\text{Log}(H^+) = 0.85$.

ii) If the acetic acid solution is ionized at only 1.4%, $pH = -\text{Log}(0.14 \times 1.4\%) = 2.7$

Q17. A 0.081 g of unknown monoacid dissolved in 1 L of water is neutralized with 6.35 mL of 0.0471 M NaOH. i) Estimate the equivalent weight of the acid. ii) Could this acid be $CH_3C_6H_4COOH$ or $CH_3CH_2C_6H_4COOH$? Explain why? The molar mass of $CH_3C_6H_4COOH = 136$ g mol^{-1} and that of $CH_3CH_2C_6H_4COOH = 150$ g mol^{-1}

Ans17. i) The reaction between the acid and the base NaOH should induce a salt and water molecules. From the given data, the number of moles of NaOH is: $(6.35 \times 10^{-3} \text{ L}) \times (0.0471 \text{ M})$ = $0.006 \times 0.0471 = 0.000283$ moles of NaOH.

Since the reaction involves the loss of one proton (monoacid), the stoichiometry of the reaction should indicate that 1 mole of NaOH reacts with 1 mole of acid. This means that the number of moles of the acid is also 0.000283 moles. Hence, the molar mass of the unknown acid is:

$\frac{0.081}{0.000283 \times 1} = 286.2$ g mol^{-1}

This mass does not match the molar mass of either $CH_3C_6H_4COOH$ or $CH_3CH_2C_6H_4COOH$.

Q18. i) Estimate the pH of a solution containing an H^+ concentration of 5.2×10^{-12} mol L^{-1}. Actually, this solution represents a detergent solution and because strong bases could attack protein structures, the detergent must show a warning label if its components form solutions with pH above 11. ii) Should the proposed detergent show a warning label?

iii) Estimate the pH values of 0.06 M ammonium chloride solution, 0.03 M sodium acetate, and 0.15 M sodium cyanide. K_a (ammonium chloride) = 5.6×10^{-10}, K_b (sodium acetate) = 5.37×10^{-10}, and K_b (sodium cyanide) = 2×10^{-5}.

Ans18. By definition, $pH = -\text{Log}(H^+) = -\text{Log}(5.2 \times 10^{-12}) = 11.28$

ii) Because the pH value is higher than 11, the box must show the warning label.

iii) The three proposed acids are all weak, meaning they only partially dissolve in water. In that case, the proton concentration of the ammonium chloride solution could be expressed by: $(H^+) = (K_a \times 0.06)^{1/2} = 5.8 \times 10^{-6}$. Thus, $pH = -\text{Log}(H^+) = 5.2$

With $K_b = 5.37 \times 10^{-10}$ and molarity of 0.03M, the hydroxyl concentration of sodium acetate solution can be estimated as: $(OH^-) = (K_b \times 0.03)^{1/2} = 5.39$. Since pH + pOH = 14, hence *pH* = (14 - 5.39) = 8.6

With $K_b = 2 \times 10^{-5}$ and 0.15M, the hydroxyl concentration in sodium cyanide is:

$(OH^-) = (K_b \times 0.15)^{1/2} = 1.7 \times 10^{-3}$, thus *pOH* = -Log (OH^-) = 2.77 or pH = (14 - 2.77) = 11.23

Q19. Equal volumes of 0.15 M acetic acid and 0.15 M sodium acetate are mixed to form a buffer solution. i) Estimate the pH of the resulting buffer solution. pK_a (acetic acid) = 4.74

A second buffer solution is prepared by mixing equal volumes of 0.3 M propionic acid with 0.3 M sodium propionate. ii) Calculate the pH of the obtained solution. pK_a (propionic acid) = 4.87.

A third buffer solution is prepared from 0.061 M ammonia and 0.181 M ammonium chloride. iii) If HCl is added to this buffer, would the pH change? Calculate the pH before and after the addition of HCl. pK_b (ammonia) = 4.75

Ans19. i) Buffer solutions are characterized by stable pH upon the addition of small amounts of acids or bases. Acetate and propionate are good species for preparing buffer solutions because of their relevant conjugate acid/base characters.

The pH of a buffer can be estimated by the expression: $pH = pK_a + Log\frac{(CH_3COONa)}{(CH_3COOH)} = pK_a + Log\frac{(0.15)}{(0.15)} = pK_a = 4.74$

ii) The same procedure applies to the propionate buffer: $pH = pK_a + Log\frac{(CH_3CH_2COONa)}{(CH_3CH_2COOH)} = pK_a + Log\frac{(0.3)}{(0.3)} = pK_a = 4.87$

iii) Unlike acetate or propionate, ammoniate has a weak buffer capacity, meaning not good for preparing buffer solutions.

Ammonia is weak base, hence *pOH* can be estimated by the formula: $pOH = pK_b + Log\frac{(salt)}{(base)} = pK_b + Log\frac{(0.181)}{(0.061)} = 5.22$. Therefore, pH = (14 - 5.22) = 8.77

Because of its weak buffer capacity, even small amounts of HCl should change the pH of the ammoniate solution. The addition of HCl should reduce a pH below 8.7 (neutralization reaction between acid and base).

Q20. A 0.11M pyridine solution is titrated by an HCl solution. Estimate the pH of the resulting solution at (equivalent H^+/equivalent pyridine) ratios of 0.4 and 1. K_b (pyridine) = 1.58×10^{-8}

Ans20. Pyridine is a weak base, and pOH can be estimated by the formula: $pOH = pK_b + Log \frac{(salt)}{(base)} = pK_b + Log \frac{(salt)}{(base)}$

At ratio of 0.4, $pOH = 7.8 + Log (0.4) = 7.4$, or $pH = (14 - 7.4) = 6.6$

At ratio of 1, $pOH = 7.8 + Log (1) = 7.8$, or $pH = (14 - 7.8) = 6.2$

Q21. Write down the equilibrium constants of each of the following reactions.

$CH_3COOH + F^- \leftrightarrow HF + CH_3COO^-$

$NH_3 + HSO_3^- \leftrightarrow SO_3^{2-} + NH_4^+$

Ans 21. Each of the given reactions can move forward (right to left) or reverse (left to right). The equilibrium constants in both directions are reversed and their ratio should give 1.

For $CH_3COOH + F^- \leftrightarrow HF + CH_3COO^-$, $K_{eq} = \frac{(HF)(CH_3COO^-)}{(CH_3COOH)(F^-)}$

For $NH_3 + HSO_3^- \leftrightarrow SO_3^{2-} + NH_4^+$, $K_{eq} = \frac{(SO_3^{-2})(NH_4^+)}{(NH_3)(HSO_3^-)}$

Q22. Silver phosphate (Ag_3PO_4) is a slightly soluble salt in water at 0.0067g L^{-1} (at 20 °C). i) Estimate the solubility constant (K_s) of Ag_3PO_4. ii) Calculate the solubility of Ag_3PO_4 in a solution containing 0.11 mol L^{-1} of Ag^+. The molar weight of $Ag_3PO_4 = 418.58$ g mol^{-1}

Ans22. i) The solubility reaction of Ag_3PO_4 can be expressed as:

$Ag_3PO_{4(s)} \Leftrightarrow 3Ag^+_{(aq)} + PO_4^{-3}{}_{(aq)}$

The equilibrium constant K_s of the reaction: $K_s = (Ag^+)^3(PO_3^{-4})$ since the concentration of Ag_3PO_4 equals to 1 (solid).

Solubility (in mol L^{-1}) is: $S = \frac{(0.0067)}{(418.58)} = 1.6 \times 10^{-5}$ mol L^{-1}

ii) The stoichiometry of the reaction indicates that 1 mole $Ag_3PO_{4(s)}$ induces 3 moles $Ag^+_{(aq)}$ and 1 mole $PO_4^{-3}{}_{(aq)}$.

Thus, $K_s = (Ag^+)^3(PO_3^{-4}) = (3 \times 1.6 \times 10^{-5})^3 \times (1.6 \times 10^{-5}) = 1.76 \times 10^{-18}$

If $(Ag^+) = 0.11$ mol L^{-1}, $(PO_4^{-3}{}_{(aq)}) = \frac{(Ag^+)}{3}$

Thus, $K_s = (Ag^+)^3(PO_3^{-4}) = (0.11)^3 \times (\frac{0.11}{3}) = 4.9 \times 10^{-5}$

Q23. AgOH is dissolved in a buffer solution at pH 12. i) Estimate its solubility in this media. Another salt ($Mg(OH)_2$) is dissolved in two aqueous solutions: one at pH 3 and another at pH 11. ii) Estimate the solubilities of this salt in both media. In your view, could the behavior of ($Mg(OH)_2$) be useful for chemical separation?

Ans23. i) The solubility reaction of AgOH in the buffer at pH= 12 can be expressed by:

AgOH → $Ag^+ + OH^-$

The solubility constant $K_s = (Ag^+)(OH^-)$ since the concentration of solid AgOH is 1.

On the other hand, $pOH = 14 - pH = 14 - 12 = 2$, hence $(OH^-) = 0.01$.

The stoichiometry of the reaction indicates that 1 mole AgOH gives 1 mole of each of the species. Hence, $(OH^-) = (Ag^+) = 0.01$

This gives a $K_s = 0.01 \times 0.01 = 0.0001 = 10^{-4}$

ii) The solubility of $Mg(OH)_2$ in the buffer solution at pH = 3 can be expressed by:

$Mg(OH)_2$ → $Mg^{2+} + 2OH^-$

The stoichiometry of the reaction indicates that 1 mole $Mg(OH)_2$ gives 1 mole Mg^{2+} and 2 moles OH^-.

The solubility constant $K_s = (Mg^{2+})(2OH^-)^2$, and $pOH = 14 - 3 = 11$

Thus, $(OH^-) = 1\times10^{-11}$ and $(Mg^{2+}) = 0.5 \times (1\times10^{-11}) = 5\times10^{-12}$

$K_s = (5\times10^{-12}) \times (1\times10^{-11})^2 = 5 \times 10^{-34}$

The K_s value is very low, indicating the very low solubility of $Mg(OH)_2$ in this medium.

Similarly, the K_s of $Mg(OH)_2$ in the buffer at pH 11 can be expressed by:

$K_s = (Mg^{2+})(2OH^-)^2$ because the concentration of the solid $Mg(OH)_2$ is 1.

The pOH= 14 - 11 = 3, meaning that $(OH^-) = 1\times10^{-3}$ and $(Mg^{2+}) = 0.5 \times 1\times10^{-3} = 5\times10^{-4}$

This gives a value of $K_s = (5\times10^{-4}) \times (1\times10^{-3})^2 = 5 \times 10^{-10}$

The solubility of $Mg(OH)_2$ in pH 3 is a lot higher than that in pH 11, suggesting that higher pH media significantly reduces solubility. This behavior could be used for separation purposes between species.

Q24. Iron reacts with steam (water in the gas phase) at $500°C$ to form hydrogen gas and Fe_3O_4. Write down the reaction and the equilibrium constant.

Ans24. At 500°C, iron reacts with steam according to the following reaction:

$Fe + H_2O$ → $Fe_3O_4 + H_2$

At the equilibrium, the formation constant can be expressed by: $K_{eq} = P_{H2}/P_{H2O}$, where P is the pressure. The activities (or concentrations) of Fe (solid) and Fe_3O_4 (solid) are 1. If water is used in excess, its concentration should also be 1. Note that the concentrations, in that case, are expressed in terms of pressure because the species are present in the gas phase.

Q25. Consider the following reaction at 298 K: $2NO_{2(g)} \Leftrightarrow N_2O_{4(g)}$, with enthalpy H = -57 kJ mol^{-1} at 298K. i) Write down the equilibrium constant of the reaction. ii) What will happen to the reaction in the following cases: increase of T, increase of P, addition of extra N_2O_4, and removal of $NO_{2(g)}$.

Ans25. The equilibrium constant of the reaction ($2NO_{2(g)} \leftrightarrow N_2O_{4(g)}$) could be written as follows: $K_{eq} = \frac{(N_2O_4)}{(NO_2)^2}$

ii) The increase in T or P should induce more collisions between the reactants to produce more $N_2O_{4(g)}$. Thus the reaction should move forward (from left to right).

The addition of extra N_2O_4 should move the reaction in the reverse direction (from right to left) to consume the excess N_2O_4 and restore the equilibrium state (Le Chatelier's principle).

Removal of NO_2 should also move the reaction from the right to left to produce more NO_2 and restore the equilibrium state.

Q26. Potassium chloride (KCl) is a highly soluble salt in water. Its solubility is estimated to 347 g L^{-1} at 20°C and 802 g L^{-1} at 100°C. Estimate the solubility constant K_s at each temperature.

Ans26. The solubility reaction of KCl in water could be summarized as follows:

$KCl \rightarrow K^+ + Cl^-$

The solubility constant K_s could be written as: $K_s = (K^+)(Cl^-)$ since the concentration of the solid KCl is 1.

At 20 °C, the solubility is 347 g L^{-1}, thus the concentration of $(K^+) = \frac{47}{39} = 8.89$ mol L^{-1} and that of $(Cl^-) = \frac{347}{35.5} = 9.77$ mol L^{-1}

This gives: $K_s = 8.89 \times 9.77 = 86.89$

Similarly, at 100 °C, the solubility is 802 g L^{-1}, hence the concentration of $(K^+) = \frac{802}{39} = 20.56$ mol L^{-1} and that of $(Cl^-) = \frac{802}{35.5} = 22.5$ mol L^{-1}

This gives a $K_s = 20.56 \times 22.5 = 464.48$

It can be seen that higher temperatures induce higher solubility and solubility constants.

Q27. Consider the spontaneous combustion reaction of liquid methanol:

$CH_3OH_{(l)} + \frac{3}{2}O_{2(g)} \Leftrightarrow CO_{2(g)} + 2H_2O_{(l)}$

In what direction should the reaction move in each of the following scenarios: i) addition of extra $CO_{2(g)}$, ii) removal of $CO_{2(g)}$, and iii) addition of $CH_3OH_{(l)}$ and $\frac{3}{2}O_{2(g)}$?

Ans27. i) The addition of extra $CO_{2(g)}$ should move the reaction from the right to left to consume the excess added $CO_2(g)$ and restore the equilibrium (Le Chatelier's principle). ii) Removal of $CO_2(g)$ should move the equilibrium from left to right to allow the formation of more $CO_2(g)$ and restore the equilibrium. iii) The addition of both reactants will move the equilibrium from left to right to produce more products.

Q28. HCl could dissociate to form hydrogen and chlorine gases, with an equilibrium constant of 6.2×10^{-54} at 25°C.

$2HCl_{(g)} \leftrightarrow H_{2(g)} + Cl_{2(g)}$

i) Write down the expression of the equilibrium constant. ii) Considering the given value of the equilibrium constant, what do you think about the dissociation of HCl?

Ans28. i) The equilibrium constant can be expressed as: $K_{eq} = \frac{(H_2)(Cl_2)}{(HCl)^2}$

ii) An equilibrium constant of 6.2×10^{-54} is very low, meaning very low concentrations of the products when compared to those of the reactants. This indicates the very low dissociation ability of HCl in this process.

Q29. i) Briefly, define the chemical potential of a substance. In what unit is it measured? ii) What would be the chemical potential of a system at constant temperature and pressure?

Ans29. Chemical potential is known as partial molar free energy, which is a form of potential energy that is absorbed or released during a chemical process or phase transition. Chemical potential is expressed in Joule kg^{-1} of the substance or Joule mol^{-1}.

ii) The chemical potential is given by the relationship: $dG = -SdT + VdP + (\mu_1 dN_1 + \mu_2 dN_2 ...)$, where dG is the infinitesimal change in Gibbs free energy, S is the entropy, V is the volume, dT and dP are the infinitesimal changes in temperature and pressure of the system, and dN_i is infinitesimal change in number of species.

At constant temperature and pressure, dT and dP equal to *zero*. Thus, $dG = (\mu_1 dN_1 + \mu_2 dN_2 ...)$, meaning that the chemical potential is directly related to changes in the free energy and number of species.

Q30. What is the relationship between the chemical potential and concentration (or activity) of species dissolved in water?

Ans30. The chemical potential of a substance dissolved in water could be expressed as:

$\mu_i = \mu_i^o + RT \ln(a_i)$, where R is the gas constant, T is the temperature, μ_i^o is the chemical potential of the species (i) at the standard conditions, and a_i is the activity of the species.

Q31. If the enthalpy of a reaction is negative and the entropy is positive, what do you think about the spontaneity of the process?

Ans31. The relationship between the enthalpy, entropy and free energy is given by:

$\Delta G = \Delta H - T\Delta S$, where H is the enthalpy, S is the entropy, and T is the temperature.

If $\Delta H<0$ and $\Delta S>0$, $\Delta G<0$. Hence, the reaction will occur spontaneously.

Q32. i) In few words, what is the activation energy of a reaction? ii) How does an enzyme influences the activation energy?

Ans32. i) The activation energy can be defined as the minimum energy required for a chemical system to induce a reaction. It is expressed in kJ mol^{-1} or kcal mol^{-1}.

ii) Enzymes are catalysts that can lower the activation energy and allow the reaction to proceed more easily at a faster rate. At the end of the reaction, the enzyme catalysts should be released to the reacting medium and separated from the products and remaining reactants.

Q33. i) Briefly, define the conductivity of an electrolyte solution. ii) What basic equipment is used to measure the conductivity of solutions? Explain the principal.

Ans33. i) The conductivity of an electrolyte solution could be defined as the formation of a flow of charge upon the passage of an electric field. In other words, when an electrolyte solution is subjected to an electric current with positive and negative polarities at both ends, the positively charged cations move towards the negatively charged pole and the negatively charged anions move to the opposite positive pole. This creates some sort of movement of charge that leads to electrical conduction.

ii) Experimentally, the ionic conductivity is measured using a conductivity meter. The basic laws linked to electricity, such as Ohm's Law ($V = IR$, with V is the voltage, I is current, and R is the resistance of the solution) also apply to the conductivity of electrolyte solutions. In general, the ionic conductivity k (S m^2 mol^{-1}) is defined by: $k = \frac{L}{RA}$, where A is the cross sectional area of the electrodes, L is the distance between the two electrodes (negative and positive poles), and R is the solution resistance. The constants L and A are often determined from calibration experiments using cells with known conductivities.

Q34. i) Explain the difference between metallic conduction and electrolytic conduction. ii) Why does the conductivity of an electrolyte increase with concentration? iii) How does dilution influence ionic conductivity?

Ans34. i) The ionic conductivity is different from the traditional conductivity occurring in metal conductors. In metal conductors, conductivity is due to the formation of electrons (charged negatively) and holes (charged positively) throughout the crystal structure. In electrolyte solutions, conductivity can be viewed as generation of flow of charge upon the passage of an electric field. In other words, when the electrolyte solution is subjected to a current with positive and negative polarities at both ends, the positively charged cations move towards the negatively charged pole and the negatively charged anions move to the opposite positive pole. This creates some sort of movement of charge that leads to electrical conduction.

ii) The conductivity increases with the electrolyte concentration because more electrical current is transported by the free ions. iii) Dilution decreases the number of free ions, thus reducing the conductivity.

Table of Content

Discount offers	1
Introduction	2
Abstract	3
1. Aqueous and non-aqueous solvents	3
2. Electrolyte solutions	3
3. Quantifying concentration of solutes	4
4. Conductivity of electrolyte solutions	5
5. Thermodynamics of electrolyte solutions	6
5.1. Energy quantities	6
5.2. Chemical equilibria	8
5.3. Some important reaction equilibria	9
5.4. Solubility/precipitation/complexation reactions	9
5.5. Acid/base reactions	10
6. Redox reactions	12
7. Kinetics of chemical reactions	12
8. Summary of important chemistry units used in electrochemistry	13
Summary	14
References	15
Practical Questions/Problems with Solutions	17
Table of content	33
About the author	35

www.ingramcontent.com/pod-product-compliance
Lightning Source LLC
Chambersburg PA
CBHW062236220526
45471CB00009B/3506